Nondual Perspectives on Quantum Physics

Tomaj Javidtash

Published in the United States of America
http://www.tomajjavidtash.com

Cover image by Nazanin Marandiz

ISBN: 1500805637
ISBN 13: 9781500805630
Library of Congress Control Number: 2014917501
CreateSpace Independent Publishing Platform
North Charleston, South Carolina

To my parents, who taught me to stand for the truth

Contents

Introduction

Quantum physics is pure mathematics enveloped in philosophical and metaphysical interpretation. The mathematical concepts of the theory—such as wave function, the superposition principle, and even the classically familiar position and momentum of a particle—have no familiar counterparts in nature as we experience and understand it. Yet quantum physics is one of the most successful and experimentally validated theories of nature, and it works just fine, despite its counterintuitive predictions.

Quantum physics is in fact an extremely sophisticated toolbox with exotic tools inside it that can tell us everything humanly possible about the microscopic realm of phenomena; it is the Swiss Army knife of atomic physics—only more sophisticated and powerful. Quantum theory is also mathematically elegant and aesthetically comparable to the Pythagoreans' vision of a musical cosmos.

What is so controversial about quantum physics is that unlike classical physics, which can give us an intuitive picture of nature, quantum physics gives us no picture whatsoever of what is *really* going on in the microscopic world. In fact, quantum mechanics tell us that the microscopic world is in principle impossible to depict or imagine. What is more perplexing is that our inability to construct a picture of this world is not due to the inadequacy of our instruments, to the incompleteness of the theory, or to human ignorance. It is not that we are simply ignorant of what is *really* happening down there. It is neither the case that we *cannot know* it because nature blocks our view; rather, we do not know because *there is nothing to know*. When we are not looking, *nothing is really going on at all*.

In the absence of observation, the microscopic realm is the state of pure potentiality: there are no distinctions, no separations, and no particle-like or wave-like phenomena. The state of the unobserved quantum world is like the state of dreamless sleep. Physicists call this state the *superposition state*, which does not exist in our ordinary space-time but in what is known as the *Hilbert space*. Particle-like and wave-like phenomena only *appear* when we poke and disturb the microscopic world, much like the disturbances that appear in a pond when a stone falls in it. Our world is such a disturbance in the infinite manifold of pure potentiality.

The underlying reality reflected in the mathematics of quantum theory is *in principle inexpressible*. The microscopic world is always in a state of pure potentially until we disturb it by our observations, as if waking up from deep sleep. Therefore, observation is in a sense creation: a waking up of the sleeping reality. What we get as a result of observation—what we create as reality—always depends on our prior intentions. If the instrument of observation is designed to poke at the microscopic world, then our poking produces sharp, pointy effects that *appear* as point-like particles. If the instrument is designed to act as an intruding wall, then our obstruction produces wave-like reactions.

The microscopic world knows no preference and segregation by itself except those imposed by us humans; it assumes no rigidity and no fixed forms; it is a world of formlessness until we invade it and impose on it our own forms and norms. This is why the quantum world cannot be depicted in principle: there are no forms to be depicted. It is we humans who create these forms as disturbances by our measuring instruments.

What we get from the microscopic world is its reaction to what we do to it. What we see in it is our own reflection; what we find in it, we ourselves have put there. In fact, what is observed in the microscopic world is produced by our looking into that world. *The phenomenal reality is a production of our intentional observation.*

It is the quantum description of underlying reality, or nonreality, as a state of pure potentiality that invokes the idea of nonduality. Astonishing similarities between the nondual philosophies of the distant past and the philosophical implications of quantum physics were the motivation behind writing *Nondual Perspectives on Quantum Physics*, book 1 of the series *Nondual Perspectives*.

There are three different human traditions that point to one underlying reality, which is essentially nondual: quantum physics in science; transcendental phenomenology of Edmund Husserl in philosophy; and Advaita Vedanta of Hindu metaphysics. It is our intention to present an accurate account of quantum physics followed by a summary of the fundamental tenets of the other two traditions; we will then show how these three traditions have arrived at the same inevitable truth, despite being historically and intellectually apart from one another.

My hope is that this work will become a humble voice among others to emphasize the fundamental unity of all beings and to invite us to contemplate this truth more often. If this unity becomes our truth and if we reflect it in all dimensions of our lives, public and private, only then can we transcend all differences and see that we are all manifestations of the same truth.

Until we see and experience this perennial truth for ourselves, our hearts are never convinced, and it is always our hearts that say the last word. I hope the time has come for our civilization to overcome all belief and conviction and instead to appeal to the direct intuition of truth so that it transforms our isolated hearts into one united heart that beats and bleeds for all beings.

"We shall see the truth, and truth shall make us free." This we shall keep with us at all times, even if all the world speaks against it.

One

Quantum Physics

The Physics

Quantum physics, quantum mechanics, and quantum theory are all used interchangeably in our discussions. It is a mathematical toolbox, a set of rules and instructions, for calculating the *probabilities* of occurrence of microscopic events. Quantum theory consists of a few postulates that introduce the necessary mathematical tools and tell us when and where they should be applied. This is what quantum theory really is; everything else is interpretation.

Different textbooks may introduce these postulates in different orders, but all physicists agree on them and follow a standard way of utilizing quantum mechanics for their calculations.

Physics, however, is the product of collective effort; it is done through engaged collaboration and communication between scientists. Therefore, physicists have to agree upon a certain way of speaking about this purely mathematical theory so that it is understood by all parties involved—such as technicians, engineers, students, etc.—in its practice and development. It is not possible to speak in mathematical terms or to give lectures that involve nothing but formulas.

In order to overcome the problem of expressing the inexpressible, physicists have to construct an *interpretation* for the theory that is expressible in natural language. When physicists discuss quantum theory in terms of an interpretation, they have a mutual agreement on the scope, the limit, and the precise meaning of its concepts. The subtleties of this interpretation are often overlooked when they are adopted by nonexperts and transmitted into the public sphere.

This careless transmission of quantum theory without emphasis on the precise meaning of its concepts is the source of much mystification of the subject followed by a flood of opportunists and fake quantum gurus that inject misunderstanding and false hope to the popular view of quantum theory. This fashionable popularization of quantum physics damages both the theory itself and those who make spiritual investments in a false version of it.

Therefore, it is very important for us to have a precise understanding of the concepts of quantum theory and to see in it only what it offers. To do so we need to use an interpretation of the theory that is legitimate and that has worked well. The most commonly accepted interpretation of quantum mechanics that is adopted by many but not all physicists is known as the *Copenhagen Interpretation,* developed between 1924 and 1927 by some of the founders of quantum theory (such as Niels Bohr, Werner Heisenberg, Max Born, etc.).

Copenhagen Interpretation is considered to be the orthodox view of quantum mechanics; thus, it is important to know the precise meaning of its concepts and their limits and to give a few examples of the common misperceptions in the popular view of the theory.

When physicists use the words *particle* and *wave,* they are aware of the fact that these words have no well-defined, let alone precise, meanings within the microscopic world. There are no point-like solid objects, neither wake-like phenomena down there. To a physicist the word *particle* implies the collection of infinite layers of *probability waves* inside an infinite dimensional abstract space that is peaked about a point. In fact, Heisenberg Uncertainty Principles, which is one of the tenets of quantum theory, asserts that a point-like particle is an impossible state of affairs. John Bell's theorem presented in his 1964 paper and validated in various experiments has shown that a *local reality*—a reality composed of isolated point-like entities—cannot produce the observable phenomena. This is what a physicist means when he utters the word *particle.*

The principle of superposition is another often misunderstood tenet of quantum physics. A consequence of this principle is that in the absence of observation, a quantum object *behaves as if* it is in many places at the same time; this state of the object is called the *superposition state*. When people say that in the quantum world a particle *is* at different places at the same time, they have misunderstood the meaning of the superposition state; because one may think that we could throw light in darkness and see a particle to be in many places at once! The first problem with this assertion is, as we mentioned before, that there is no such thing as a point-like particle. We may at best find a point-like trace of a quantum particle left on the sensors of the measuring instrument. The second problem is that the superposition state of a quantum system is its state *in the absence of observation*. The moment we make an observation of this system, its *superposition state collapses* into a point in our familiar three-dimensional space.

An electron is never *seen* or *observed* to be at many places. The superposition state is really just a mathematical expression of the physical state of a quantum system when it is not interacting with its environment: that is, before the system acquires a specific form. *Superposition state* is a mathematical expression of the state of *pure potentiality*. What is potential is not yet actual. Therefore, a particle in *superposition state* is not yet an actual object, let alone is it actually present in many places at the same time.

Experimental data from the microscopic world have shown that quantum systems behaved as if they possessed no fixed forms or objective properties except mass and charge prior to and in the absence of observation. By objective properties, I mean those properties that exist in the object independently of our knowledge of them. But quantum systems do not seem to have concrete, localized existence when they are left to themselves; they are in and of themselves formless.

All empirical evidence points to the formlessness of quantum objects prior to the act of measurement. In fact, according to the principles and predictions of quantum theory, if a quantum particle *did possess* objective properties such as position (location in 3-D space) and momentum (mass × velocity), then we would not observe phenomena such as interference patterns or entangled states of quantum systems; but we do observe these phenomena.

In spite of the fact that prior to the act of measurement a quantum particle does not possess objective properties such as position or momentum, the measurements of position and momentum always yield a numerical value for its position and momentum! In other words, although the quantum particle is quite formless prior to being subject to measurement, it is always measured to have one form or another. Measurement of an electron's position always gives us a value for its position; and measurement of an electron's momentum always gives us a value for its momentum.

As space and time are a priori *forms* of sensible intuition in Kantian terms—no perception can occur outside space and time—position and momentum, too, are a priori forms of all natural *phenomena*. By phenomena I mean the *appearance* of things: things as they are being observed or experienced. Thus, we cannot experience nature without its elements *appearing* to have a position and momentum with our experience. It is easy to see how a priori forms of natural phenomena position and momentum are the extensions of Kant's a priori forms of sensible intuition space and time: position rests on the concept of space, and momentum rests on the concept of time and change of position in time.

A major difference between quantum physics and classical physics is that in classical physics it was assumed that particles possessed a precise position and momentum, whether or not we measured them; but quantum physics showed that *it was the act of observation that produced properties such as position and momentum in the particle*.

In fact, even during the very short period of observation it is not really the quantum particle that acquires a position or a momentum; position and momentum are effects belonging to the *interface of interaction*, the place where the interaction between the quantum particle and the measuring apparatus takes place.

It was in expressing this peculiarity of quantum systems (formless entities always being observed in a form) that physicists had to invoke the *principle of superposition*, saying that prior to measurement there is only potentiality. To see how this principle is the only reasonable way to mathematically express the peculiarity of the quantum world, let us briefly restate these peculiarities:

A quantum particle does not have any specific numerical value for an observable quantity (like position or momentum) when it is not being observed; but whenever it is observed, it picks a specific numerical value from the spectrum of all possible values of that observable. For instance, all measurements of the position of a particle always yield a numerical value for its position; but the particle did not have a specific position prior to measurement!

But there is even more: assume that we have exact replicas of the same particle, and we toss them in space with exactly the same initial conditions (in the same direction and with the same speed). Now, if we simultaneously measure the position of all of these particles, we find each to be in a difference place! This is in violation of the laws of classical physics, saying that the same initial conditions must give also the same objective paths for a collection of identical particles.

In short: 1) The quantum particle has no position right before the measurement of its position. 2) The quantum particle always *appears* to have a position right after the measurement of its position. 3) The particle takes this position, jumping from potentiality into actuality, instantaneously and randomly.

An object that does not *actually* possess anything but always gives us something on demand is a kind of object that *potentially* possesses everything; such an object whose essence, the "in itself," is a state of pure potentiality, is no more a physical reality but a metaphysical ideality. The *principle of superposition*, or the *superposition state*, is the mathematical way of writing down an equation for this kind of metaphysical construct.

My choice of the word *metaphysical* by no means implies that the state expressed by using the principle of superposition is unreal. On the contrary: this state is very real, but it is not physical. Reality in itself is not physical or material but metaphysical and ideal; the underlying reality is pure potentiality. What we *perceive* as physical reality with apparent material constitution is nothing but the manifestation of an ideal, nondual reality that pervades all phenomena. What *is*, is real and nondual; what is perceived is mere appearance and has no material constitution behind it.

A particle in a superposition state is not a material object in many places at once; rather, it is in a state of pure potentiality: that is, it has no actual position. The concept of position, or momentum, does not apply to an object spending its time in potentially; in the same way that the concept of place no more applies to us when in the state of dreamless sleep.

The superposition state, which is constructed from the principle of superposition, describes the state of a quantum system in the absence of observation—the state of pure potentiality. Each time we make a measurement of a property of this system, the *state* of the system is changed; our measuring apparatus acts on the system and changes its form.

In quantum physics, there is no such thing as *passive observation,* meaning an act of observing something without disturbing what is being observed. The reason is that at the microscopic scale, the objects under investigation are the same size as the light particles by which observation is made possible. To observe something, at any scale, means to shine light on it and see the reflection of that light. When light is reflected from the surface of an apple so that we can see it, the apple does not roll back as a result of the impact of light, because the particles of light are too weak to push the apple. But if we wanted to do the same thing with an electron instead of an apple, the electron would recoil and move away. Imagine trying to bounce a basketball off another basketball! That is why an observation, a measurement, or any interaction with a quantum system changes the state of the system.

The quantum system is always characterized by a *quantum state,* which contains all that we can know about the system at a particular point in space and time. For instance, when an electron goes from point A to point B, its state has changed, though it may have the same velocity as before or even not have a well-defined velocity; this is itself part of the information contained in the state.

A state of a quantum system is mathematically characterized by an abstract mathematical object called *Wave Function* or *State Vector*. The same way that even numbers are members of a larger set containing all numbers, wave functions are members of a larger set called *vectors,* belonging to an abstract mathematical space called the *Hilbert Space.* In short, a wave function that represents a state (possible form) of a quantum system is an abstract mathematical object dwelling in Hilbert Space.

One of the mysteries of quantum physics lies in the fact that wave functions themselves do not correspond to any familiar element of physical reality; their only connection to reality is that they can be mathematically manipulated into another class of objects that give us the probability of the occurrence of a microscopic event, such as the probability of an electron being at point A or point B.

Much like Newton's equations that determine the state of motion of a classical system, there is in quantum theory one equation that determines the evolution in time of the wave function of a quantum system: *Schrodinger Equation.* Schrodinger Equation tells us how the wave function representing the state of the quantum system changes over time. For instance, assume that we have a free electron roaming the room; its wave function tells us that the probability of this electron being found at point A in the room is 30 percent. If we solve the Schrodinger Equation for this wave function, the solution will tell us what the wave function of the electron will look like at a later time. From that we can calculate the probability of the same electron being found at point A, or any other point, in a future time.

Notice that quantum mechanics can only tell us the *probability* of the occurrence of a microscopic event. All we can have is a wave function; and all that a wave function can tell us is the probability of a quantum system being found in such and such a state when it is observed. Quantum physics does not tell us with certainty what the exact state of a quantum system will be at some point in space and time.

According to the orthodox view of quantum physics, Copenhagen Interpretation, our lack of knowledge about the exact state of a quantum system is a fundamental aspect of the theory, and it cannot be overcome.

The fact that we do not know the exact state of a quantum system is not because of our ignorance; rather, it is because *there is nothing to know*. It is not that the information about the exact state of the system is existent, but that it is shielded from us; such information is nonexistent to begin with. The quantum system *does not have an exact physical state until the moment that it is observed.*

Before the act of observation, the system is in a *superposition state* that *potentially* contains all the possible states of the system at once. It is the very act of measurement that *destroys* the superposition state and forces the system to leave the state of potentiality and enter the state of actuality by *quantum jumping* into a particular, actual state from among all of its possibilities available in the superposition state. The transition from the *superposition state* to an actual state, called an *eigenstate*, which occurs as a result of measurement, is an instantaneous event known as *the collapse of the wave function*. More precisely, during the measurement process, the wave function in the *superposition state collapses into an eigenstate*: from potentiality into actuality.

Another important tenet of quantum theory is the *Uncertainty Principle,* first derived by Werner Heisenberg. Every measurement involves some uncertainty, however small. According to the Uncertainty Principle, there exist properties of a quantum system such that the uncertainty in the measurement of one property is inversely proportional to the uncertainty in the measurement of another. We cannot know everything we want to know about a quantum system; these systems like to keep us in the dark.

Modern quantum mechanics, developed in the 1920s, was formulated in two separate attempts: first, Werner Heisenberg formulated the *Matrix Mechanics* version of it in 1925. In his version, quantum states and observables, measureable properties, were represented by mathematical objects called matrices.

Later, Erwin Schrodinger formulated the *Wave Mechanics* version in 1925–26. In his version, quantum states and observables were represented by waves and differential operators, respectively. However, Schrodinger's waves were no ordinary waves; they were *probability waves* in an infinite dimensional

Hilbert Space. The wave function of a quantum system of which we spoke above is Schrodinger's probability wave.

Erwin Schrodinger later showed that these two versions were different representations of the same theory: Quantum Mechanics.

In his formulation of Matrix Mechanics, Heisenberg realized that in some cases when he changed the order of two matrices, the results of their multiplication were different. This is a peculiarity of matrices that does not arise in basic algebra: 2 multiplied by 3 gives the same number as 3 multiplied by 2. But matrices representing quantum states and observables turned out to violate this basic rule known as *commutativity*. Further study of these matrices and their relations to quantum states, observables, and measurements led Heisenberg to the formulation of his infamous *Uncertainty Principle*.

In the quantum world, there are physical properties of a quantum system called *Complementary Variables* such that a simultaneous knowledge of them in the same system and with arbitrary precision is impossible. These complementary variables appear in pairs: such as position and momentum, energy and time, etc.

The Uncertainty Principle states that we cannot know the exact position and momentum of a particle at once because the uncertainty in the measurement of the position of a particle is inversely proportional to the uncertainty in the measurement of its momentum. The same relation holds for any pair of complementary variables. Thus, the Uncertainty Principle is about the *intrinsic relation* between uncertainties involved in measurements; no such relation was observed to exist in classical physics. The following is a more accurate picture of an uncertainty relation.

Say that you measure the position of an electron and get a numerical value for its position with some uncertainty. Now you want to know the momentum of the same electron. You have to perform a momentum measurement, which involves a different apparatus. Say that you perform this measurement and get a numerical value for its momentum with minimized uncertainty. According to the Uncertainty Principle, your measurement of momentum destroys your information about the exact position of the electron obtained in the first measurement. The degree to which the uncertainty in momentum measurement is minimized, the uncertainty in position measurement is maximized.

If you measure the position of that same electron after the momentum measurement, you will get a totally different value than the one in the first position measurement. This third position measurement itself destroys your previous information about the electron's momentum.

I like to compare this situation to making cookies with different shapes using a cookie cutter. Imagine that you have two cookie cutters of different shapes: a circle and a square. First you make the dough into a square-shaped cookie using the square cookie cutter. Now, if you apply the circle cookie cutter on the same piece of dough, you ruin the previous shape and instead get a circle-shaped cookie. If you had just softly touched the dough with the circle cookie cutter without pressing it too hard you would have gotten a cookie with both traces of a square and a circle. The quantum system reacts in a similar way to our instruments: the underlying reality is a quantum dough.

We see that the order of measurements of complementary variables makes a great difference in the outcome. In the case of a single quantum particle, if you measure the position first and the momentum second, you will end up with an exact value for momentum but nothing for position. Measuring the momentum first and the position second, you will end up with an exact value for position but nothing for momentum. This is the gist of the Uncertainty Principle, and no more mystery should be read into it.

The correlation between uncertainties exists only between complementary variables. There are variables that are not complementary; the measurement of one of these variables does not destroy the information about the other. These are called *compatible variables*, such as momentum and energy. For instance, if you measure the momentum of a particle and right after that measure its energy with perfect precision, then your measurement of energy did not destroy your information about its momentum. You can know the momentum and energy of a particle at the same time and with arbitrary precision; hence, the compatibility.

This was a simplistic approach to quantum mechanics. There are of course more approaches, but almost all of them are built upon the basic postulates and principles that we reviewed here. Most of what we see in a quantum mechanics textbook are highly sophisticated mathematical tools for dealing with complex systems and interactions. What is of interest to us in this study are the meanings and philosophical implications of these principles that point to an unfamiliar underlying reality.

The Philosophy

Quantum physics is a description of the microscopic world. Classical physics is a description of the macroscopic world. These different realms of phenomena must be aspects of the same world, only in different scales. It is then a logical matter that these two descriptions should not be at odds with one another. But it is an experimental fact that they are, at least apparently.

The world depicted by classical physics is a familiar world. It is a description of the world of human perception. Our common sense, logic, and understanding are rooted in our experience of this world. That is why the theories and models of classical physics are intuitive, even though their mathematics may be difficult. The classical picture of the world is essentially based on human experience and intuition. And the sophisticated mathematics used in classical physics has a purely utilitarian value: it is a mere language reflecting the regularities of a material world.

But when we enter into the microscopic world, all intuitions break down. Since our logic and understanding are grounded in our experience and since we have no direct experience of the microscopic world, there is no way for us to truly understand it. There is no way to imagine or depict this world, because imagination and depiction always draw on our previous experiences, which are nonexistent in the case of microscopic world.

Quantum physics is a mathematical toolbox for calculating probabilities; but unlike classical physics, whose mathematics is only a language for describing the material world of experience, there is no familiar, material world underneath the mathematics of quantum physics. Mathematics is all there is in quantum physics: it is both the tool and the underlying reality. That is why many physicists, including Werner Heisenberg, have entertained the idea that the microscopic world is indeed a mathematical world: a world of mathematical ideas and constructs.

The classical, materialistic picture of the world has collapsed under the pressure of experimental evidence based on which quantum theory was developed. If the strange ideal mathematical reality described by quantum physics reminds you of Plato's world of ideas, you are not alone. Werner Heisenberg, one of the founders of quantum physics, believed that the reality described by modern physics was more Platonic than anything else. He wrote in his *Physics and Frontiers*:

> *I think that on this point* [the nature of reality] *modern physics has definitely decided for Plato. For the smallest units of matter are in*

> *fact not physical objects in the ordinary sense of the word; they are forms, structures or—in Plato's sense—Ideas, which can be unambiguously spoken of only in the language of mathematics.*

A world with such ideal mathematical infrastructure is not familiar to our classical intuition. Advances in experimental physics and its ability in further exploring the microscopic world and verifying the predictions of quantum theory have proved again and again that our world is essentially quantum mechanical, and that classical physics and its pictures are only approximations to a quantum mechanical world.

The reason we do not perceive quantum mechanical effects in everyday life is that these effects are noticeable only in very small scales and for system with smaller degrees of freedom. But almost all of the large-scale phenomena, except for gravitation, are consequences of the laws of quantum theory. These laws dictate the behavior of all microscopic as well as macroscopic phenomena.

Since our world is an essentially quantum mechanical world, the reality underlying all phenomena is no more understood to be a physical, material substratum but an undifferentiated state of potentiality; the *material appearance* of our world is only a manifestation of this undifferentiated substratum. It should be noted that the hypothesis of the material constitution of the world has always been a metaphysical assumption. If we are to remain objective and empirical, we cannot deny that we always only *experience* a material reality and never actually encounter it without the mediacy of our subjective experience.

The underlying reality that is exposed in quantum mechanics is a state of pure potentiality that knows no distinctions: no here or there, no this or that; it is a nonlocal, universal unity that only *appears* to be diverse, differentiated, and local as a result of becoming conscious of itself and observing itself. After all, who is it that is making the observation and thus forcing the wave function to collapse into an actuality?

If we view the whole situation from a wider and more objective perspective, we realize that the observer, the observing instrument, and the act of observation are all parts of the same closed system—the universe.

The observing agent whose observation *creates actuality by destroying potentiality* is himself or herself inside a larger quantum system—the universe. He or she is made of the same elementary particles, obeying the laws of quantum physics. To be more precise: observation is something done by some elementary particles on some other elementary particles using a bunch of other elementary particles. This whole process involves nothing but the universe itself. Humans and their science are not outside the universe; they are but potentialities *of and inside the universe*. Ironically, this universe that contains humans and their science is itself a product of a science itself produced by humans!

The information we gain about a quantum system by observing it is really an information gained by the universe itself. It is the universe that knows this universe *as a universe*. If we define universe as the largest quantum system and define consciousness as *awareness of information*, then we can restate the above as follows:

It is the universe that is being conscious of itself and creating itself through its own self-consciousness by shooting itself out of pure potentiality into actuality in what is known as the *collapse of the wave function*.

All this happens inside one universe. Isn't it obvious that our universe is really *our* universe—the universe as known to humans through human reflections and human inquires about the human conscious experience of an all too human world? All these reflections and inquiries are themselves *acts of consciousness*!

This universe is the universe *seen* and *thought*, the universe *inquired and experimented* upon, the universe *known* and *advertised*, and the universe *theorized* and *believed*. There is no aspect of this universe that is not shaped in and through human acts of consciousness. There is nothing that is known without the absolute and inevitable mediacy of consciousness, because to know is to *experience* knowledge.

The universe we know is a universe we are in, thinking of it as the universe. In other words, this universe that we think contains us is something already contained in our consciousness as an object of its knowledge. It is always in and through consciousness that we are aware of this universe as a universe containing our awareness of it. The brute fact we come to is that there is no universe just existing there in itself; *there exists only our consciousness of a universe*. We can never pass beyond this consciousness without using consciousness itself.

An objective analysis of the logical philosophical implications of quantum physics tells us that the observer and its knowledge are parts of the same closed system we call the universe, which is itself an object of knowledge of the observer's consciousness.

It is time to come to the enigma of all enigmas: first, remember that a quantum system's wave function naturally dwells in pure potentiality and is actualized when the system interacts (observation is a kind of interaction) with something outside itself, an environment. In the absence of any interaction and when the system is perfectly isolated, it always remains in a superposition state—a state of pure potentiality.

We also know from quantum mechanics that each quantum system, regardless of the number of its components, can be assigned a single wave function that contains all the information about the system as a whole. We may define our quantum system to contain one electron or one billion electrons; in either case, one wave function is enough to describe the entire system. Quantum mechanics treats its objects, whether an atom or a galaxy, as a whole and not as composed of separate parts. There seems to be an intrinsic connection between the seemingly separate components of a whole of which quantum mechanics is aware.

This way of treating a system always as a whole is one of the consequences of the principle of superposition. In fact, wave functions are kinds of entities that add up like colors: if you have ten different oil colors and mix them up perfectly, you will end up with one color and not ten or seven. Colors mix up in such a way to yield *one* new color; they *superimpose* to form a unity. Waves are the same: white light is one color, yet it contains within it the seven different colors of the rainbow. These subcolors already exist in the whiteness as potentialities; a prism can actualize and differentiate these potentialities as different colors.

Wave function acts exactly the same as regular waves when it comes to combining with one another: a billion electrons forming an isolated, untouched quantum system mix up in such a way that their identities and distinctions become the potentialities of the system as a whole; the whole system is then assigned one single wave function.

Now let's do this: since any quantum system is given one wave function that collapses only upon interaction with an environment, what would happen if we defined our system to include the whole environment without leaving anything out? That is, what if we defined our quantum system to be the whole of the universe?

We assign to this whole system a single wave function representing the largest possible universe that contains our universe. This is all in agreement with quantum physics. Here is the enigma of all enigmas:

The wave function of this universe can never collapse. Due to the way we defined the system, there exists nothing outside it so as to interact with it, forcing its wave function to collapse into an actuality. This universe is always in a superposition state; it remains forever in a state of pure potentiality. To sum it up: *our universe cannot possibly be real; it cannot have come into existence. Nothing has ever happened.*

This assertion is entirely a consequence of the laws of quantum theory. According to quantum physics, our universe, or the largest universe containing ours, cannot happen unless of course we put a consciousness outside this totality. But this move does not interest us here due to the purely metaphysical and unverifiable nature of the assumption.

What is even more disturbing is that we know about the impossibility of our universe while being inside it! This conclusion is a piece of information that exists inside of a supposedly unreal universe. It is as if the universe both exists and does not exist!

This side effect of quantum physics—the universe both existing and not—is the enigma of all enigmas; and there is no way out of it unless quantum physics is false. But we know that quantum theory is the most successful and the most experimentally validated science in human history!

If this state of affairs gives you a headache, it did not do so for ancient mystics, particularly those of the Hindu tradition of Advaita Vedanta. In fact, their sages and mystics have frequently expressed exactly the same proposition as the one we concluded from the principles of quantum physics: *this world is but is not real*. And they came to this conclusion a few thousand years ago.

This hypothesis of the unreality of our universe is now being taken even more seriously in cosmology. Many cosmologists are considering the possibility that perhaps our universe is a *holographic universe*. Our universe could be a holographic projection played on a three-dimensional spherical surface enclosing a four-dimensional singularity. In this picture, what is in fact *real* is the central singularity, while the Big Bang and our universe are as *unreal* as the events in a film that is projected on a screen: they are *mere appearances* as good as our dreams, only more stable.

According to ancient mystics of Hindu and some other traditions, our universe is a *mere appearance*, an illusion that is only *experienced* as a reality. They claimed to have seen this truth during their mystical experiences of higher states of consciousness. Drawing upon their mystical intuitions, they concluded that in the same way that a dream is the experience of an objective-looking reality that has never happened, our universe is a mere subjective experience whose reality is created not by an actually existing material world but by the experiencing consciousness itself.

They called this "experiencing consciousness" the *witness consciousness*. They experienced it as a dense and singular point of pure consciousness whose *outward projection* creates the illusion of the world when rays of consciousness are refracted by the mind.

These mystics have often described higher states of consciousness as states of pure potentiality in which there are no distinctions and no divisions between the subject, object, and knowledge. Their descriptions remind me of the superposition state of which our universe is a mere possibility—something not yet happened.

Mandukya Upanishad is one of the most philosophical Vedanta texts of Hinduism. Verse seven of its first section gives an account of the state of the true Self, a state called *Turiya*; this Self is seen as pure witness, unaffected by and disinterested in the world of appearance. Everything that is, is in Turiya:

> *They* [sages] *consider the Fourth* [quarter/state] *to be that which is not conscious of the internal world, nor conscious of the external world, nor conscious of both the worlds, nor a mass of consciousness, nor conscious, nor unconscious; which is unseen, beyond empirical dealings, beyond the grasp (of the organs of action), uninferable, unthinkable, indescribable; whose valid proof consists in the single belief in the*

> *Self; in which all phenomena cease; and which is unchanging;, auspicious, and nondual. That is the Self, and That is to be known.*

It is the nondual tendencies inherent in the philosophical foundations of quantum physics that suggest the resurrection of perennial truths of nondual traditions. As serious seekers of the highest truths, we must overcome ideological preferences and fashionable convictions, whether theistic or atheistic, and value only objective inquiry. In this spirit, we shall suspend all opinion and proceed in the sacred path of intuition.

Two

Nonduality

The Idea

It has been a traditional practice among our species to try to *understand, to know* the answer to the world's first and last question: "*What is this?*" referring to the whole of this existence; all the other big questions come later. Almost all such traditions trying to answer this question sought to find a unity within the apparent diversity.

Humans like to find a principle or a set of principles that explain everything; in their pursuit of knowledge, they have always been after something underlying the world of phenomena: the world as experienced by them. The origin of the idea that there may be something more to this phenomenal world—let alone one or a few knowable principles—could be in the phenomenon of dreaming or in the experiences of higher states of consciousness reported by mystics.

This tradition of looking for a valuable vintage in the basement of reality took different turns in the course of its evolution. One thing that we are sure of is that that human's conception of the underlying principles moved toward a unity through higher levels of abstraction. These principles were first identified with the manifest forces of nature; then there was polytheism and its anthropomorphic gods who ruled nature but were still tangible and exciting. Then grew monotheism and beard, and the principle of interest became a unique and all-pervading god, beyond nature, but yet a being who spoke a few languages and was very much interested in the world's affairs.

Metaphysicians interested in logic and rigor took this unique being to the unreachable heights of intellectual abstraction; they posited that the principle had to be *Absolute* and *Infinite*: "To say more of it," they argued, "is to limit it." Thus, the metaphysician's god was no more a tangible being but rather a *beyond being*; it could no more be imagined or called upon. Though this level of abstraction did not find any fans among the interest-driven masses, it became the favorite anchor point for the speculative philosophies of Aristotle and the Scholastics.

At the verge of the scientific revolution of the sixteenth century, Galileo Galilei took up the unfinished Greek project of *rational science* and invigorated it with the ready-made Euclidean Geometry: he put mathematics at the foundation of nature and pronounced them man and wife. Galileo established a new tradition based on a purely quantitative approach to reality; he transformed the underlying principles from metaphysical-intellectual to physical-mathematical. All of modern science, about which we proudly brag, is just one human tradition, among others, based on one dogma: mathematical representation of nature corresponds to nature itself. Quantum mechanics is indeed the Messiah of the Galilean tradition of science. It has come to end and transcend this tradition of purely quantitative treatment of man and nature.

Despite apparent differences and even contradictions, all of these traditions in pursuit of understanding were motivated by the idea that the diversity of the phenomenal world follows from a few basic principles, whether godlike or mathematical. Nonduality was a guiding idea in some of these traditions.

Of the traditions in pursuit of truth, we consider nondual those that 1) posit *one* unchanging principle governing all phenomena and 2) consider the phenomenal world to be one with that principle in such a way that the principle is not separate from its manifestation, the world.

In the sense defined above, monotheism fails to be a nondual tradition because monotheistic religions, from the exoteric point of view, conceive of the one principle (god) and its manifestation (creation) as separate entities, though one is the cause of the other. This is not the case in nondual traditions, in which any distinction between the principle and its manifestation, whether logically or ontologically, is considered a contradiction in terms.

The most dominant dualistic tradition goes back to Rene Descartes, seventeenth-century French philosopher and mathematician, who postulated that the fabric of reality consists of two distinct and yet cooperative substances: mind and body. He divided all phenomena into the physical and the mental. We modern humans are all synthetic Cartesians with a touch of disregard for the mental and a great regard for the physical.

The nondualists, unlike Cartesian dualists, believe that the one underlying principle always preserves its unity; it is never divided into two or more; it has always been nondual and will always remain so. Thus, the nondualist position does not admit creationist accounts of existence, rendering the world an illusion.

I have to add that despite the apparent nonduality in monotheistic religions, the esoteric forms of these religions and their practitioners mostly adhere to a nondual interpretation of its exoteric forms. They subscribe to one truth underlying all religious and spiritual traditions. In fact, this underlying nondual principle, *The Perennial Wisdom*, is something that almost all mystics agree upon, regardless of their traditional backgrounds. The source of such universal agreement is in their mystical experiences, *intuitions* of a truth transcendent to its various formal expressions.

Despite the grand intellectual-metaphysical edifices attached to most esoteric traditions with the aim of inferring the unity of all existence from first principles, esoteric truth itself is not transmitted by means of logical demonstration or metaphysical argument but rather through direct, intuitive seeing of the truth. Their metaphysical constructs are only articulations of what is already seen and realized with absolute certainty.

However, the nonduality that interests us here with regard to quantum physics is one which is expressed from the outset, both esoterically and exoterically. Our interest in nonduality arose from the philosophical implications of quantum physics: the underlying reality, the state of superposition, is essentially a nondual state of affairs composed of potentiality rather than actuality.

We also saw that applying the principles of quantum theory to the universe itself brings us to the apparently illogical conclusion that *our universe has never really happened*! Yet this conclusion is itself a piece of knowledge experienced inside an apparently existing universe! We considered this paradoxical situation as the enigma of all enigmas.

But there existed a nondual branch of Hinduism known as *Advaita Vedanta* that had come to exactly the same conclusion about the universe, with a slight difference in reaction. We see it as enigma; they saw it as common sense.

Advaita Vedanta

Advaita Vedanta is the nondual subdivision of Hindu philosophy. Hindu philosophy, which is based on the teachings of the *Vedas* (sacred texts revealed to sages) yields both dual and nondual interpretations. Vedas, meaning knowledge, were revealed and assimilated during a wide span of time from a few thousand years ago to a few hundred years before Christ.

The last portion of the Vedas, which are highly philosophical texts, are known as Vedanta, meaning the end of the Vedas: the conclusion of knowledge. The main texts of Vedanta include *Upanishads*, *Bhagavad Gita*, and *Brahma Sutras*.

Advaita means *not two*; therefore, Advaita Vedanta literally means *Nondual Vedanta*, *The Nondual Conclusion of Knowledge*. It is the philosophical assertions of Advaita Vedanta that have direct relevance to our discussion, particularly those expounded by *Shankaracharya*, the eighth-century Hindu philosopher who actually established Advaita Vedanta as an authentic Hindu doctrine grounded in the Vedas.

It was some years after I had taken my undergraduate courses in quantum mechanics and had become comfortable with its abstract mathematics and ideas that I came across one of the Upanishads; I saw mention of it in Martin Heidegger's *Discourse on Thinking*. I had picked up the Kena Upanishad. I was so fascinated by the similarity of its ideas to the underlying reality of quantum physics that I was convinced they were equivalent but alternative descriptions of the same reality.

It is very important not to confuse Advaita Vedanta with religious dogma: Advaita Vedanta is a highly scientific and sophisticated account of conscious experience.

Its truths are phenomenological truths established first by the direct experiencing of higher states of consciousness and then by logical and philosophical analysis based on empirical evidence. Its method of inquiry is *phenomenological,* namely: pure description of lived experience as it is given to consciousness.

Instead of relying on the analyses of a third person who has only indirect access to my states of consciousness, I objectively observe and study these states as they are being experienced by me. This is the prominent method of investigation in the phenomenological approach to consciousness.

Advaita Vedanta, too, is a phenomenological approach to reality with the difference that it is not limited to a narrow range of human experiences recognized only by the majority. It also incorporates into its body of knowledge the intuitive experiences of higher states of consciousness attained by yogis and mystics. Herein lies the power of its objectivity.

Two concepts from Hindu philosophy's account of creation are importance for us because they have counterparts in quantum theory: *Purusha* and *Prakriti.* Although these concepts existed in Hindu cosmology prior to its nondual phase, Advaita Vedanta absorbs them into itself as separate aspects of the same nondual essence—pure consciousness. From the point of view of Advaita Vedanta, these concepts are more symbolic than literal. In reality, *pure consciousness* is all that there is; it is an unchanging state in which no creation has ever happened.

Purusha is the active, masculine agent of creation, and *Prakriti* is the passive, feminine agent of creation. These two agents alone are considered to be responsible for the appearance of objective reality. According to Hindu cosmology, the phenomenal world springs from the interaction of these two agents.

In Hindu cosmology, *Purusha* signifies the rays of consciousness that produce the appearance of the universe by acting upon *Prakriti,* which is an undifferentiated state of pure potentiality. *Prakriti*, the Primordial Nature, is the realm of the *Unmanifest*. When *Purusha* acts upon *Prakriti,* what is potential is projected outward as real and as an actual *experience of the universe*. What is actualized depends on the preexisting ideas inherent in *Purusha*. It is *Purusha*'s intention that is actualized and experienced by itself when its rays of consciousness act upon the state of pure potentiality, *Prakriti*. At the same time, in virtue of the Absolute and Infinite nature of pure consciousness, no change is possible in it. The actualization of potentiality, though it exists as appearance, itself remains forever a potentiality within the infinite consciousness. The world-appearance is nothing but a daydream for the unchanging pure consciousness.

The constitution of the reality described above is very similar to the constitution of the reality expressed in the mathematics of quantum theory. The superposition state, which is a state of pure potentiality in the abstract *Hilbert Space,* corresponds to the concept of *Prakriti,* which is the passive state of pure potentiality existing outside space-time, because it contains space-time as one of its possibilities. The action of observation/measurement that actualizes one particular state out of pure potentiality is similar to the way *Purusha* actualizes an appearance by acting upon *Prakriti*.

Another significant similarity is how the outcome of the actualization already depends on the constitution of the actualizing agent. In Hindu cosmology, what is actualized from the background of pure potentiality, *Prakriti*, is determined by the ideas already existing in *Purusha*. In quantum physics, the outcome of observation heavily depends on the constitution of the measuring apparatus, which in its own part is rooted in human intentions behind the design and application of the apparatus.

We observe particle-like phenomena because the measuring apparatus is designed to produce particle-like effects by poking the quantum system. The similarity between the form of the eigenfunction (which is the wave function of a quantum system immediately after being observed) and the actual process of measurement indicates this dependence, which always finds its way back to the intentions of the scientist.

I would like to make one last point before delving into the precious words of Advaita Vedanta. It is true that performing a measurement on a quantum system throws that system out of the superposition state and into an actual state—eigenstate. It is also true that this procedure does not require the presence of a conscious human person at the moment of measurement.

Whether we choose to look at the result of measurement or not, the measurement has been performed, and the wave function has collapsed. But we should note that the very act of measurement and the constitution of the measuring apparatus is already marked by human intentions and involvements. Human consciousness is always indirectly present through its representative, the measurement apparatus.

In order to see the status of our proposed enigma within the philosophical framework of Advaita Vedanta, we quote from various Vedic sources some phrases with direct relevance to our discussion of quantum physics. Please keep in mind the symbolic import of what is said in these quotations. According to Hindu sages, these writings are illustrations and allusive indications designed to convey to the spiritual aspirant an essentially nondual and inexpressible reality that cannot be grasped by the dualistic mind; and it is the aspirant who is the primary audience of these texts.

Chapter 3, verses 17 and 18 of Mandukya Upanishad:

> [17] *The dualists, confirmed believers in the methodologies establishing their own conclusions, are at loggerheads with one another. But this (nondual) view has no conflict with them.*
>
> [18] *Nonduality is the highest Reality, since duality is said to be a product of it. But for them* [nondualists] *there is duality either way. Therefore this view (of ours) does not clash (with theirs.)*

That duality is the product of nonduality is established for the mystic in higher states of consciousness known as *Samadhi*; thus, for them it is considered to be an empirical fact. Verse 31 of the same chapter states:

> *All this that there is—together with all that moves or does not move—is perceived by the mind (and therefore all this is but the mind); for when the mind ceases to be the mind, duality is no longer perceived.*

This verse refers to the logical relation between the object and the awareness of object. An object insofar as it is something known, something to be known, or known as something inaccessible to our knowledge derives its objectivity from first being an object of awareness; awareness always precedes its object.

Throughout all its cognitive modifications (as perceived, as thought, as theorized, etc.), the object is first and foremost an object of awareness; it has no separate existence from the awareness through which it is intended as such. The object is something essentially cognized in one way or another. Even the inaccessible thing in itself is an object of awareness, precisely as *the thing in itself*; it is an object of awareness insofar as it is a product of ideation.

What creates the impression of independent existence of an object is its *name*; but name is just a tag that stands for the manifold of *subjective* experiences through which an identity, the *idea* of object, is intended by consciousness. In fact, if one takes away from an object its various subjective manners of appearance, nothing remains of it but a mental idea and an empty name.

An object is nothing but a mental constitution synthesized within consciousness throughout its various appearances. A phenomenological analysis of dreams confirms the same position held by both the Vedantist and the Phenomenologist. Verse 37 of chapter 4 of Mandukya Upanishad states:

> *Since a dream is experienced like the waking state, the former is held to be the result of the latter. In reality, however, the waking state is admitted to be true for that dreamer alone, it being the cause of his dream.*

The idea presented in this verse is very important in our comparison of Advaita Vedanta and Transcendental Phenomenology to be discussed in the next section. According to the Vedantist, the source of reality or unreality is in the beliefs held within consciousness. The apparent reality of the world is an idea whose origin is in the constitution of conscious experience.

It is precisely due to the structural features of consciousness that the dream world is experienced as a real world when we are in it. When we wake up, the waking state becomes subject to these internal structures of experience; hence, we wake up to this world with a predetermined belief in its reality.

Although we have dreamt thousands of times, each time being convinced of its unreality upon waking up, our next dream is still as real as the first, and the dream world appears as objective as the waking world: the phenomenon of objectivity belongs in the internal structures of subjectivity rather than experienced contents.

For the same reason that only waking up from a dream can convince us of its unreality, we can't help but believe in the reality of our waking state until we are convinced otherwise; that is, until we wake up from it. The experience of *Samadhi* in Hinduism and the method of *Transcendental Reduction* in phenomenology are practical methods of waking up to the unreality of this seemingly real world. It is only then that our naïve, however unconscious, belief in the existence and reality of this world is irreversibly destroyed; and it is only then that we see for ourselves that the real objective appearance of this world was caused by strong *beliefs* embedded within the structures of our natural experience of the world.

It is to be noted that the unreality of this world and the possibility of waking up from it into an undifferentiated state of pure awareness is not a metaphysical construct; it is a simple matter of direct experience to see it and become convinced of its truth. That is why no proof of it was ever necessary: for its proof is its sight. What should motivate the unbiased mind in the direction of seeing it for himself or herself is the fact that it is impossible to prove the existence and reality of this world.

Verse 51 of section 4 of the same Upanishad states:

> *When Consciousness is in vibration, the appearances do not come to It from anywhere else. Neither do they go anywhere else from Consciousness when It is at rest, nor do they (then) enter into It.*

This verse and the following verses from the same section have relevance to our discussion in quantum physics in the fact that our universe appears to exist but is not supposed to be in existence:

> [55] *Cause and effect spring into being so long as there is mental preoccupation with cause and effect. There is no origination of cause and effect when the engrossment with cause and effect becomes attenuated.*
>
> [57] *Everything seems to be born because of the empirical outlook; therefore there is nothing that is eternal. From the standpoint of Reality, everything is the birthless Self; therefore there is no such thing as annihilation.*
>
> [58] *The entities that are born thus are not born in reality. Their birth is as that of a thing through Maya (magic.) And that Maya again has no reality.*

> [61] *As in dream Consciousness vibrates as though having dual functions, so in the waking state Consciousness vibrates as though with two facets.*
>
> [73] *That which exists because of a fancied empirical outlook, does not do so from the standpoint of absolute Reality. Anything that may exist on the strength of the empirical outlook, engendered by other systems of thought, does not really exist.*
>
> [79] *Since owing to the belief in the existence of unrealities, Consciousness engages Itself in things that are equally so (i.e. unreal); therefore when one has the realization of the absence of objects, Consciousness becomes unattached and turns back.*

After Upanishads, we come to the most exalted of nondual philosophical texts: *Yoga Vasistha*. It was written by the Hindu sage Valmiki, speculated to have lived between 100 and 500 BC. *Yoga Vasistha* is in the form of a dialogue between sage Vasistha and Rama, a Hindu deity. Below I quote from part four, *On Existence*, of a translated edition by Swami Venkatesananda.

> *This world-appearance is experienced only like a day-dream; it is essentially unreal. It is a painting on void like the colors of a rainbow. It is like a widespread fog; when you try to grasp it, it is*

> *nothing. Some philosophers treat this as inert substance or void or the aggregate of atoms. Creation is just a word, without corresponding substantial reality.*

Also, from *The Philosophical Verses of Yogavasistha*, translated by Swami Bhaskarananda:

> *Any attempt to see or know Maya* [Cosmic Illusion] *causes its destruction. Therefore, the nature of Maya cannot be seen or known.* [This is similar to the phenomenon of the collapse of the wave function. This reminds one of the destruction of interference patterns in Young's double slit experiment; any attempt at seeing the paths of particles causes the destruction of the patterns.]
>
> *Due to ignorance about the Atman* [supreme Self] *the world appears to exist. It ceases to exist when the knowledge of the Atman is acquired.*
>
> *There is nothing like a cause and effect relationship. Everything appears to exist due to erroneous thinking. Experience of the existence of everything arises from Brahman who is Consciousness itself.*
>
> *Know the mind to be this world. It has been said that the mind itself is the bondage. Just as trees are swayed by the wind, so also this body is moved by the mind.*

These sayings reflect the general idea behind the metaphysics of Advaita Vedanta. The following is a summary of this idea:

There exists a nondual reality known as Brahman, which is none other than pure consciousness itself. Brahman is the only thing that exists. The logical categories of existence, space, and time do not apply to Brahman because Brahman itself is the condition for the possibility of existence, space, and time. Brahman is the Absolute and the Infinite. Everything that *appears* to exist is but a manifestation of Brahman. Manifestation is neither existent nor nonexistent. It has the ontological status of a dream. It is mere appearance: the self-projection of Brahman. Manifestation has *relative existence* and *absolute nonexistence*. Like a dream that exists only as dream but does not exist as reality, this universe too is nothing but an appearance of existence produced by the mind—the mind itself being an illusion within the infinite consciousness. Nothing has ever been created; existence is mere illusion.

The conclusion of Advaita Vedanta's metaphysics is similar to our conclusion from quantum mechanics: both claiming that our universe seems to exist, though it is not supposed to. I should emphasize that quantum mechanics has no intention of being metaphysical; neither did Advaita Vedanta have any intention of being a physical theory of nature.

Quantum mechanics as a physical theory was invented to account for the experimental facts of *subatomic world*; and Advaita Vedanta as a metaphysical doctrine was established to express in nondual terms the truth behind the inexpressible mystical experiences of enlightened sages. Our comparisons are meant to show that the philosophical implications of quantum mechanics from a nondual perspective have many parallels in Advaita Vedanta.

We now turn to what is known as *Transcendental Phenomenology,* which is a philosophical movement of the twentieth century. Phenomenology is not directly associated with nonduality, but it has come to conclusions almost exactly like those of Advaita Vedanta.

Transcendental Phenomenology

> *Perhaps it will even become manifest that the total phenomenological attitude and the epoché belonging to it are destined in essence to effect, at first, a complete personal transformation, comparable in the beginning to a religious conversion, which then, however, over and above this, bears within itself the significance of the greatest existential transformation which is assigned as a task to mankind as such.*
>
> —Edmund Husserl

Edmund Husserl, a twentieth-century German mathematician and philosopher, is the founder of Transcendental Phenomenology, also referred to as *Phenomenology*.

He was the teacher and mentor of German phenomenologist Martin Heidegger.

Husserl's phenomenology was a philosophy unlike all others because it set its starting point of inquiry to be the direct, lived experience of the philosopher's surrounding world. Seeing how the unjustified metaphysical assumptions underlying sciences and philosophies of his time had thrown out of its correct course the project of a genuine, objective science, he set out to "return to things themselves," to pure phenomena as they were given to consciousness. He began the inquiry from the very beginning—namely, immediate conscious experience.

Edmund Husserl's phenomenology did, hopefully, to the whole of Western philosophy precisely what quantum physics did to classical physics. Phenomenology's emphasis on the *return to phenomena themselves* was a call for a paradigm shift also invoked by quantum theory's emphasis on the *return to observables themselves.*

In order to be rigorous and scientific, Husserl had to discard all the assumptions and presuppositions that were injected by science and philosophy into our conscious experience of the world. One such metaphysical assumption underlying all sciences, according to Husserl, was taking for granted the reality of the experienced world.

We know of this world in and through conscious experience; all our inquiries, our scientific experiments, our theorizing, and our philosophizing are deliberate acts of consciousness. Consciousness is the accomplishing agent behind all these acts in and through which we come to the world and know it as such.

The claim that the world known through these deliberate acts has an existence independent of the very consciousness by which it is known is an assumption that cannot be verified at all. Any attempt at verifying the independent existence of the world, insofar as it becomes a conscious involvement in the world, destroys the very independence under investigation.

Independent existence means existence in the absence of our conscious involvement in the world. But all such verifying attempts involve an observation of some kind, which brings back into the system some sort of an involvement, hence, destroying its independence.

It was precisely the problematic consequences of this metaphysical assumption that were exposed and then eliminated by quantum theory. Quantum physics shook our classical view of the world by asserting that it is meaningless to speak of a predetermined reality that exists independently of our interactions with it.

Edmund Husserl refused to keep this *reality assumption* in his project of finding a firm ground for a genuine science. He initiated his inquiries into the foundations of the world of experience by suspending all beliefs in the reality of the experienced world. He also refused to favor the opposite and assume that the world was nonexistent; instead, he set out his phenomenological inquiries such that their results would be independent of whether the world existed or not. He began from *pure phenomena*.

After explaining in detail the metaphysical nature and the problematic consequences of the *reality assumption*, Edmund Husserl articulated the project of phenomenology in his book *Ideas; General Introduction to Pure Phenomenology*:

> *No proofs drawn from the empirical consideration of the world can be conceived which would assure us with absolute certainty of the world's existence.*
>
> *Our phenomenology should be a theory of essential Being, dealing not with real, but with transcendentally reduced phenomena.*

There does not have to exist a real world in order for consciousness to have an experience of a real world. We all know one such case: dreams. The dream world is pure phenomenon with no substantial reality. Yet, our experience of *seeing* in a dream is the same as our experience of *seeing* in the waking state: both experiences have the same formal structures, although we believe one of them is not real.

In both cases, objects are seen in perspective; they have an apparent shape and size; and they always appear with some sides facing us, while some others are facing away and out of view. We can find these structural similarities in all modes of cognition by comparing their functions in a dream to those in the waking state.

Conscious experiences have certain internal structures that are always present and functioning whether or not the contents of those experiences actually exist. Phenomenology is the study of these internal, logical structures of experience that function independently of the reality or unreality of the experienced world. These internal structures characterize the conscious acts through which our world is experienced and known. For instance, due to these internal, logical structures, the experience of seeing is always immediately recognized as seeing and not as hearing or touching. Seeing has a self-evident character for the seer.

According to Edmund Husserl, we came to our experience and consciousness of the world in virtue of these internal structures of experience. The world that we believe contains our consciousness is a world itself known in consciousness.

The *sense of reality* of the world is an ideal, *non-empirical component* present in all moments of experience because it springs from the internal, logical structures of consciousness itself. Sense data may represent phenomena, but they cannot convey *realism*.

Edmund Husserl realized that this *sense of reality* of the world, also present in dreams, is a product of the logical structures of experience itself and not an actual attribute of the world. The *sense of* reality and existence of the world always taken for granted is itself a mere phenomenon, a subjective tail, and a product of consciousness.

The internal structures of consciousness *constitute* our experiences with an accompanying *mood of existence*. It is this mood wrapped around all our experiences that we interpret as *existence*. The *sense of existence* is just a secondary quality of the world, as opposed to primary qualities, embedded in all subjective experiences. It is there in our experience as the sensation of color is there in experience. That is why even our dream experiences, and any experience whatsoever, are always equipped with an unquestionable belief in the reality of what is experienced.

Quantum physics came to a similar conclusion a few decades after Edmund Husserl. According to quantum theory, reality is created, out of potentiality, in and through observation. A measuring apparatus upon performing its measurement throws a quantum system into an actual state. A yet unobserved reality has no meaning.

In order to keep his phenomenological analysis unaffected by *natural presumptions,* Edmund Husserl devised a method of inquiry by which he could enter into and stay in a particular mode of experiencing that made *pure phenomenology* possible. Husserl called this particular mode of experiencing the *Phenomenological Attitude*. This manner of experiencing stands in sharp contrast with what Husserl called the *Natural Attitude,* comprised of our natural, human ways of experiencing the world.

Everything we do as humans happens inside the natural attitude. In fact, we are not even aware of this attitude as an attitude because we have never been outside it. Our natural way of experiencing the world in which we find ourselves as *humans inside a real world* is just one way of experiencing—a particular attitude of consciousness—and we have taken this attitude for granted as if it were the absolute reality.

The phenomenological attitude, however, is an attitude by which we can for the first time experience the world as pure phenomenon. In this particular mode of consciousness, we no more experience ourselves and the world as existent entities; we no more see ourselves as human beings inside a real world. In phenomenological attitude, the world is no more experienced as extended beyond this particular here and now.

From this new angle, we see for the first time that there is nothing outside the now moment: there is no world, no human being, and no past and future. The world is a purely *subjective* phenomenon that only *appears* to extend beyond the here and now; it only *appears* to have a history and a future. Naturally, we fail to *see* this because we are too caught up in this thing called our humanity and deeply captivated by its involvements in the world.

Natural attitude is the realm of *natural experience*; phenomenological attitude is the realm of *transcendental experience*.

The method by which one can shift consciousness from its natural attitude to its phenomenological attitude is given various names by Edmund Husserl: *Phenomenological Reduction, Transcendental Reduction, Transcendental-Phenomenological Reduction, Universal Epoché, Bracketing,* or *Reduction.*

Phenomenological Reduction involves a suspending of all our natural beliefs in the reality of the world. These unconscious beliefs are always at play underneath all our experiences. In performing the *Reduction,* the phenomenologist *puts out of play in one blow,* to use Husserl's phrase, all beliefs and assumptions attached to his or her natural experience of the world.

In the case of successfully performing the Reduction and entering into the realm of transcendental experience, you come to the most shocking moment of existence in the face of which you lose your tongue: you have fallen unto the ground of all experience; you have come face to face with the naked, inexpressible truth, a truth previously clothed by your own ideas and beliefs.

The "I" that is left of you in *the ground of experience* is no more a person; it is not an object. It is a pure, *experiencing* consciousness. We then realize that our natural experience had no real subject to begin with; the only thing that there is, is *pure experiencing awareness.* The "I" that you experience is no more a thing; it is a verb: it is the act of experiencing. This "I" is also called the *Phenomenological Onlooker,* which is similar to Vedanta's concept of *Witness Consciousness.*

The manner of existence of this new "I" is entirely different from that of natural objects; it is something beyond existence because time is no more experienced the way we know it. Using the language of quantum theory, one can say that the *Transcendental Experience* of the world is the *ground state of experience* upon which all the rest is built as layers of belief.

It was to this astonishing and yet shocking experience that Edmund Husserl referred in the paragraph quoted at the beginning of this section. The quote was taken from page 137 of his book *The Crisis of European Sciences and Transcendental Phenomenology—An Introduction to Phenomenological Philosophy*.

The message of Husserl's phenomenology and the truth experienced and established during *transcendental experience* activated by performing the *Reduction* can be summarized as following:

The only existing reality is pure consciousness, a living present; it is not inside something nor outside anything. It is the ground of space and time, hence, itself beyond space and time. The world and our humanity are constitutions of this one underlying consciousness, all-pervading field of awareness, which is the common ground of all experiencing subjects. There exists only one consciousness, and different subjects are mere appearances experienced by this consciousness. This is done in the same way that during a dream, a person's consciousness *appears* both as him or her and as a world with other subjects in it.

Subjectivity and intersubjectivity are accomplishments of the same underlying consciousness. The world is really nothing; it has no separate existence nor a substantial reality; it is pure phenomenon, a mere appearance. More precisely: *the world is nothing but a moment of the transcendental experiencing consciousness.*

The underlying pure consciousness, which is identical with the experiencing "I" that is left after the Reduction, is eternal; it is not born and does not die. The moment that Reduction is performed, we know this truth with absolute certainty. We instantly know it because we experience time differently and without any *outsidedness* to it; time has no outside; it has no past or future. Time *appears* to have a past and a future only in the experience of our natural mode.

In reality, when *transcendentally reduced,* time is no more experienced as having a before or an after; it has not come from anywhere and is not going anywhere either. This new *sense of time* and *temporality* is understood only through direct experience rather than imagination. Before stepping into the realm of transcendental experience, it is impossible to fully grasp what it is like to experience everything from the outside.

Our natural mode of experiencing time, which I call *ecstatic time* because of its outsidedness, is the root cause of all our existential questions, including the questions about the origin and future of our universe. From the transcendental standpoint, all these questions are meaningless, because nothing is really happening, and time is not really passing.

In transcendental experience and due to the character of *transcendental time,* questions of origin and teleology do not arise to begin with. The truth of the matter is that time has no before or after; we have not come from anywhere, and we are going nowhere. There is really nowhere to go; the *onlooking I* is all of it.

This pure experiencing consciousness is all that there is; the instant we do the Reduction and experience it for ourselves, all our questions melt away. These questions arise in the first place because of our ecstatic experience of time. No more questions arise after the Reduction, because we have realized that there is nothing to know; there is nothing to do; there is nowhere to be; nothing has ever happened or will ever happen, because there is nothing to happen. There is only experience: a sophisticated, panoramic, 3-D daydream of pure consciousness; and we are that: the immortal experiencing consciousness.

Transcendental phenomenology goes beyond all objectivity and subjectivity; the truth disclosed by the Reduction method is essentially a nondual reality, transcending all object-subject distinctions.

Here I would like to quote once again a short sentence from *Yoga Vasistha* that expresses the same thing as does Husserl about the source of the sense of reality of the world:

Experience of the existence of everything arises from Brahman who is Consciousness itself.

Edmund Husserl's Transcendental Phenomenology through empirical analysis and its Reduction method came to the same truths as those claimed in Advaita Vedanta metaphysics. In fact, they are identical formulations, though historically and culturally apart, of the same underlying reality: pure experiencing consciousness.

In Vedanta, the realization of truth was an experiential fact: something to be experienced directly. All sages in guiding the spiritual aspirant toward the realization of truth emphasized one thing over and above everything else: letting go of the belief in the reality of the world. This was the core of their path toward enlightenment and liberation. In fact, the central doctrine of Advaita Vedanta about the source of the illusion of the world is that it is our own beliefs in the reality of this world that make it appear as real; it is the power of consciousness (a power known as Maya) that makes it experience as real whatever it believes to be real.

This method of *letting go* of the belief in the reality of the world as a means of attaining enlightenment is identical to Husserl's Reduction as a means of entering the realm of pure experience. Sage Vasistha frequently reminded Rama that the way to liberation is the cessation of the belief in the reality of the world:

> *Rama, he who has realized the unreality of material substances sees only the one undivided consciousness everywhere.*
>
> *Liberation is the realization of the total non-existence of the universe as such. This is different from a mere denial of the existence of the ego and the universe.* [Husserl warned students against the same possible misunderstanding about Reduction.] *The latter is only half-knowledge. Liberation is to realize that all this is pure consciousness.*
>
> *Even as the mirage appears to be a very real river, this creation appears to be entirely real. And, as long as one clings to the notion of the reality of "you" and "I," there is no liberation.*
>
> *Bondage consists in the belief that the visible world is real, and release depends on the negation of the phenomenal.* [This would be a Vedantic restatement of Reduction.]

The only difference on this point between phenomenology and Vedanta was that Hindu sages did not know the exact mechanism behind such *letting go* or *Reduction*; but Edmund Husserl discovered the precise mechanism by which one came face to face with pure consciousness.

The experience that mystics have always spoken of as enlightenment is none other than the *ground state of experience* in which the absolute reality of pure consciousness and the nonexistence of the manifest world are unveiled. What happens as a result of the Reduction method is in fact a very sudden waking up from the dream of this world; it is a freeing of the captivated gaze of pure consciousness; it is the losing of whatever you are and everything you know; it is the instant of realizing that you are at once the seeker, the sought, and the way.

We conclude our presentation of Transcendental Phenomenology by quoting another paragraph from Husserl's book *Ideas I,* about the apparent reality of the world and its absolute dependence on consciousness:

> *Reality, that of the thing taken singly as also that of the whole world, essentially lacks independence. Reality is not in itself something absolute;* [it is] *binding itself to another only in a secondary way; it is, absolutely speaking, nothing at all; it has no "absolute essence" whatsoever; it has the essentiality of something which in principle is only intentional* [intended by consciousness], *only known, consciously presented as an appearance.*

Three

Unity

Common Ground

We saw that the three intellectual traditions of quantum physics, Advaita Vedanta, and Transcendental Phenomenology all point to a nondual reality underlying all phenomena. Although each of these traditions has its own concept and name for this reality, all of them agree that no human concept can fully capture its essence.

Concepts work by differentiating and confining their objects; but the nondual underlying reality of which these three traditions speak is an undifferentiated state of affairs; thus, it is by definition immune to all conceptualizations.

Mathematics, however, is a different story. The infallible strength and beauty of mathematics, its infinite generosity, and its superiority to everything human make it the lord of our universe. If music is the magic of the heart, mathematics is the magic of the intellect.

Hans Christian Andersen has said of the transcendent power of music:

Where words fail, music speaks.

We say of the transcendent power of mathematics:

Where concepts fail, mathematics prevails.

The only possible common ground that can express the inexpressible in these traditions has to be mathematics. We have to seek a mathematical structure within which one principle can have infinite manifestations without undermining the unity and integrity of the principle itself. What is of utmost importance is that the manifestations should entirely depend on the principle, while the principle should not depend on anything but itself. The manifestations should enjoy only a dependent and *relative* ontological status, while the principle should be logically and ontologically self-sufficient and absolute.

Fortunately in mathematics there exist such a kind of objects that can stand for our inexpressible principle: vectors. The relationship between a vector and its components conveys precisely the idea of an absolute principle and its relative manifestations. The common ground we are seeking should be based on the language of *Vector Algebra*.

We know that quantum physics is already in possession of a mathematical formalism that expresses a nondual reality. Therefore, we should be able to use the mathematical formalism of quantum theory to express the nondual reality of the other two traditions. To our surprise, the mathematics of quantum theory is also based on the language of vector algebra.

The mathematical formalism of quantum mechanics is a linear algebraic structure involving vectors and infinite dimensional spaces. This formalism was adopted and developed by physicists because it was the only suitable candidate for expressing the nonintuitive experimental facts of the microscopic world.

We will see that there is no candidate better than *linear vector algebra* for expressing the Advaita Vedanta metaphysics. The reason lies, as we said earlier, in the way a vector is related to its components.

To gain a true understanding of quantum physics, it is crucial that we become familiar with the meaning behind its mathematics. We will see later that this meaning behind mathematics is precisely the same mystical-philosophical meaning present in the nondual reality expressed in Advaita Vedanta and Phenomenology. This meaning is our common ground.

Our discussion of the mathematical formalism of quantum physics does not actually involve any mathematics. Since we are interested in the meaning behind the mathematics, it will be sufficient only to develop an understanding based on familiar analogies.

Common Language

Vector is a mathematical object; it is a self-sufficient unity very much like a number. As there are rules and operations for numbers, there are also rules and operations governing vectors. As we can represent a number in different number systems, we can also represent a vector in different vector systems; these vector systems are called *Basis* or *Vector Spaces.* As numbers have digits, vectors have *components.* Since the digits of a number depend on the number system in which it is represented, the *components* of a vector depend on the *Basis* in which it is represented. So vector is really just a name for a more general idea of number. If we interpret number as a zero dimensional object, then vector is its one dimensional counterpart.

We often speak of a vector in terms of its representation in some *basis. Basis* is just a technical term for the *sign system* within which a vector is represented. We can represent a vector in an infinite number of *bases* as we can express one idea in an infinite number of languages. In order to make this subject—and quantum physics with it—more tangible, I would like to introduce a very useful analogy by which a *vector* is compared to an *idea* and a *basis* to a *language system.* This analogy would cover the whole of quantum physics.

Let us define an *idea* as an abstract object in the mind that is always independent of the language in which the idea is expressed. The idea of an apple is the same idea in the minds of all people who know what an apple is; but the linguistic expression of this same idea varies in different languages.

An idea is a single entity, an integral unity; but its expression, the word standing for the idea, may consist of many *components*: different letters. For instance, the idea of an apple is represented in English by the five-letter word *apple*. The same idea is represented in French by the five-letter word *pomme*. It is represented in Farsi in the three-letter word *sib*; while it is represented in Spanish in the seven-letter word *manzana*.

The idea itself has no components, no letters; it belongs to the mind but not to any particular language; yet its representation is always made of some components. The number of possible representations of an idea depends on the number of language systems we can use to express it.

A *Vector* compares to an idea. A *Basis,* which is the system in which the vector is represented, compares to a language. If we treat each idea as an abstract vector, then each language is a possible basis for the representation of that idea. As each representation of an idea in a language consists of some signs (letters) and a specific order of putting them next to one another, each representation of a vector consists of some ordered elements, called its *Components*.

Components of a representation of a vector are like letters of a word. Another possibility in language that has a counterpart in vector algebra is translation. Translation of a word involves taking the representation of the word in a language and representing it in another language. Through translation, the meaning of the word, the idea behind it, remains the same, while its linguistic representation changes. There is a similar procedure in vector algebra known as *Transformation*; it changes the representation of one vector in a basis into its representation in another basis. As translation takes the idea from one language to another, transformation takes the vector from one basis to another.

The fourfold structure *Idea–Language–Word–Translation* is equivalent to the fourfold structure *Vector–Basis–Representation–Transformation*. This is pretty much the gist of the vector algebra at the foundation of quantum physics.

In quantum physics, each particle, a quantum system, is assigned a vector known as the *state vector*; this state vector is represented by the wave function of the quantum system. It contains all the information about the state of the quantum system. Observation, or measurement, of a quantum system corresponds to creating a *representation* of its state vector in some *basis*. Thus, the act of observing a particle in quantum mechanics compares to the act of expressing a word in a language!

The characteristics of a basis (representational system) in which the state vector (quantum object) is represented are determined by the constitution of the measuring apparatus. If the apparatus that is acting upon the system, thus pushing it out of superposition (potentiality), is for the position measurement, then we say that we are working in the *position basis*. If the apparatus acting upon the system is for the momentum or the energy measurement, then we say that we are working in the *momentum basis* or the *energy basis*.

We said that the measurement process corresponds to selecting a basis and representing the state vector of the quantum systems in that basis. In vector algebra, the act of representing a vector in a basis is given a name: *Projection*. We then say that a vector is projected onto a basis, which is the same as saying that we selected a basis and represented the vector in that basis. The component representation of a vector in a basis is also called the vector's *projection* in that basis. The five-letter word *apple*, the representation of the idea of apple in English, is also said to be the idea's *projection* in English.

In vector algebra, a *basis* for representation is also called a *space*. To use our analogy, English language is a space in which ideas can be represented using the English alphabet. Every language, insofar as it is a representational system, is also a linguistic space in which linguistic objects—words—exist. This is in accordance with our intuitive notion of space as a place in which objects exist. Our familiar three-dimensional space is really a position space, because it is a place for objects with a location. The space of color is where colors are defined as its objects. A vector space is a space whose objects are defined to be vectors.

Every language has a finite number of *distinct* letters, its alphabets, for constructing words that represent ideas. English has twenty-six letters. Farsi has thirty-two letters. Russian has thirty-three letters, etc. These letters constitute the *maximum* number of distinct basic units in a language that can be used to represent an idea.

Likewise, in vector algebra, every *basis* or *space* has a maximum number of distinct basic units that can be used to represent a vector in that basis. These basic units of a *space* are called its *basis vectors.* Basis vectors are the alphabets of a vector space.

The maximum number of distinct *basis vectors* of a space is called the *dimension* of that space. Using this terminology, we can say that English language is a twenty-six-dimensional language, while Russian language is a thirty-three-dimensional language, etc.

In order to develop a better intuition of the concepts of vector algebra, hence grasping the meaning behind quantum physics and Vedanta metaphysics, we use these concepts in the following example involving an idea and its representation:

We say that the word *apple* is the projection of the vector, idea, "apple" in the twenty-six-dimensional space of the English language. This space has twenty-six *basis vectors,* distinct letters, for representing an idea. We chose only four* of these basis vectors to represent the idea "apple." The selected letters (*A, P, L, E*) are the *components* of the idea "apple" in English language.

Since each of these basis vectors, letters, constitutes one dimension of our space, and since we chose only four letters to represent the idea "apple," we can say that the word *apple* is a four-dimensional object in a twenty-six-dimensional vector space. Our vector space here is a linguistic space.

**Although the word apple has five letters, one of them is used twice; there are only four distinct letters in this word; since basis vectors refer to distinct units of representation we say that the idea "apple" is projected onto four basis vectors, namely:* A, P, L, E.

• • •

We see that ideas are independent of their representations. Ideas exist on their own. Representations, however, are dependent upon the ideas behind them. The existence of a representation depends on the existence of its idea. An idea exists whether we represent it or not; but a representation is no more a representation if no idea exists behind it.

An idea can be represented in many ways, but it always remains independent of any particular representation of it. Words are *represented ideas,* while ideas themselves exist beyond words. An idea is not made of letters and has no dimension; letters and dimensions apply only to representations. A word always belongs to some language, while the idea itself does not belong to any particular language.

All conceptual categories—such as projection, dimension, component, basis vectors, etc.—apply only to representations and not to ideas themselves.

An idea exists independently of all these categories. An idea has *absolute, independent existence,* while its representation has *relative, dependent existence.* I emphasize this point because it is the origin of the strangeness of quantum physics.

Our analogy is now complete. We saw that there exists a one-to-one correspondence between the concepts of vector algebra used in quantum theory and the intuitive concepts used in language when we express ideas. The following table contains all the concepts of vector algebra and their corresponding concepts in language.

Vector Algebra		**Language**
Vector	→	*Idea*
Space (Basis)	→	*Language*
Projection	→	*Expression*
Representation	→	*Word*
Basis Vectors (unit vectors)	→	*Alphabet (units of expression)*
Dimension	→	*Number of Letters of Alphabet*
Components (of a vector)	→	*Distinct Letters (of a word)*
Transformation (between bases)	→	*Translation (between languages)*

We now have at our disposal all the necessary concepts for grasping the meaning behind quantum physics. Concepts developed through our analogy will help us understand why the quantum world is so strange.

Through these concepts, we can see that the same underlying structure that makes the quantum world so strange lies also at the foundations of Advaita Vedanta and Phenomenology: the underlying reality in all three of them has the same *vector-like* character that can be best described in the language of vectors.

We will learn that the undifferentiated reality behind all phenomena is like a vector whose existence does not depend on anything outside itself. The world that we experience, the universe as we know it, is not the ultimate reality; it is only the projection, representation, of a principle, an *idea*, beyond the phenomenal world.

What we observe is a mere projection—an appearance; it has the ontological status of components of a vector. It only has a relative, dependent, existence, but the projected principle is itself absolute and always transcendent to its manifestations, the same way that a vector is always above and beyond its many representations.

The one truth that quantum physics, Advaita Vedanta metaphysics, and Transcendental Phenomenology, each in its own way, have come to realize is this:

Phenomenal reality is the projection of the Absolute, Infinite, Pure Consciousness.

Quantum Reality

An isolated quantum system is always in a superposition state, a state of potentiality, before it is observed. It has no inherent characteristics except charge and mass. Position and velocity do not apply to a quantum system prior to measurement, because such properties are produced as a result of the measurement.

Isolated quantum objects, such as an electron, behave like ideas: they exist in an abstract mathematical space. What is observed of these abstract objects during the measurement process is their *expressions/projections* in our measurement apparatus; it is the apparatus that plays the role of a *basis* in quantum physics.

The expression we get on the apparatus is the representation of the quantum object in terms of the capabilities of the measurement apparatus. If the apparatus measures the position of quantum objects, then by using this apparatus on an electron, we obtain the *representation* of our electron in the *position space*. Thus, the position we assign to an electron is only its expression in our apparatus and not an objective property of the electron itself—in the same way that the letters of the word *apple* do not belong to the idea "apple."

As it is meaningless to assign letters to an idea, it is equally meaningless to assign a position, or a velocity, to an electron in itself. This is the point of departure for quantum physics. According to quantum theory, what was previously thought of as objective reality has no real, independent existence in itself; it is a mere expression in the observer who plays the role of a *representational basis.*

Philosophers divided the qualities of an object into *primary qualities* and *secondary qualities.*

Primary qualities of an object are defined to be those that are independent of any observer; whether the object is lost in deep space or it is being observed, these primary qualities are not affected because they exist in the object itself. Examples are shape, size, and according to classical physics, also the state of motion of the object, which includes position and velocity.

Secondary qualities of an object are defined to be those that depend on the observer; these qualities exist whenever the object is being observed or experienced in some fashion. In other words, secondary qualities exist in our experience of the object and not in the object itself. Examples are color, taste, smell, apparent size and shape, etc. Color, for instance, does not exist in the object; the object is really made up of atoms and electrons that are colorless and tasteless. What we experience as color comes from our brain's interpretation of the frequencies of the light emitted by the atoms of the object.

Although secondary qualities do not exist in the object itself, the object has to be present for us to experience its secondary qualities. These qualities, therefore, are neither in the object nor in the subject;

instead they belong to, and exist in, the *interface of interaction*. They only appear during the *event* in which subject interacts with object.

What happened in the advent of quantum physics was that the primary, objective qualities of our world turned out to be only secondary qualities. Based on experimental findings, quantum physics showed that qualities such as position, velocity, energy, and hence the state of motion of an object do not really exist in the object itself; they belong in the *interface of interaction,* which is now the measurement apparatus, namely, the representational basis. Quantum physics undermined the idea of *Realism*.

This paradigm shift undermining the objectivity of our world is what makes quantum physics counter-intuitive. Once we accept that an isolated quantum object possesses no specific location or velocity or energy, then we can't even imagine such an isolated system—not so much because we do not know what it looks like, but rather because a quantum object is not a kind of object that *looks like* at all, let alone looking like something. The quantum object has no look, and hence, it is not something to look at!

The properties of quantum objects, and those of the objective reality that is essentially quantum mechanical, exist only during observation and experience. The world exists only in our *experience.* Belief in the independent existence of the world is itself something inside experience.

The world does not possess any objective property, because the world is not a kind of thing to possess anything whatsoever. The world is just an appearance.

Quantum particles—such as electrons, protons, and all other subatomic particles—are only mathematical constructs, like Plato's regular solids. A measurement apparatus displays only some digits on it, which are later to be interpreted by us humans as something meaningful.

The world of phenomena, the world we know, is not something of which we have experiences; it is not something that exists apart from experience; it exists only *within* observation and experience. What we perceive is not the world as an entity but what appears on the *interface of interaction*. Werner Heisenberg writes in his *Physics and Philosophy*:

> *We have to remember that what we observe is not nature herself, but nature exposed to our method of questioning.*

More than being just a description of subatomic world, quantum physics is a statement about the classical picture of the world. It exposes the inconsistencies and paradoxes inherent in our everyday view of the world as something material, deterministic, and indifferent to consciousness. For instance, *Heisenberg's Uncertainty Principle* rejects the possibility of material solidity.

Now let us see how the superstitious ideas of solidity and material constitution originated in the first place and why they are false.

The classical view of the world is rooted in our conscious experience. We *experience* objects as solid, liquid, or gaseous; thus, we have these main distinctions in classical physics.

Objects of our experience appear to enjoy independent existence subject to certain causal relationships, hence, the objective and deterministic aspects of the classical view. But if we pay careful attention to the very experiences within which these ideas arose in the first place, we realize the following:

All objects, and in fact the whole of our world, are first *given* to us in experience; but experience is a *temporal* stream of consciousness; the solid appearance of objects is itself something *experienced* in time. Solidity is itself a phenomenon that belongs *in* the temporal flow of experience. (If I don't use the phrase *my experience,* it is because the "I" is itself something experienced; it belongs in the content of experience and not outside it. Experience is not something *I have* because I am myself something experienced.)

The world and its objects are phenomena constituted *within* the temporal flow of subjective experience. It is *in* time that they are what they are. Objects are first and foremost *temporal entities*, for it is in and through time, as flow, that their existence has any sense. In other words, objects are first extended *in* time before they are extended in space. It is within the primordial river of time that reality makes its first appearance and acquires its objectivity.

It is important to distinguish between something that flows and something that exists *within* the flow. The flow of a river is the flow *of* water molecules; these molecules exist even when they are not flowing as the river; flow is only one possible state of motion for them.

However, the temporal flow of experience within which objective reality is constituted is essentially different from the flow in the sense of a river. Temporal flow is a state of constant flux within which *everything* is constituted; nothing exists outside this flux because existence is itself something constituted *in* time.

The fact that reality is essentially a temporal phenomenon is not calculated into our classical view of the reality. The Uncertainty Principle in quantum theory is a technical restatement of the nonphysical but temporal constitution of phenomena. This is why a point-like particle makes no sense in this primordial flow that we call the *Heraclitean Flux*.

An analogy can help us see how something can exist only *within* a flow: imagine that you are seeing something that appears to be a circle from afar; when you get close to it, you see a bright circular object hanging in space. Although there appears to exist a circular object, the bright circle is nothing but the trace of a ball of fire that is being rotated very fast by an invisible fire dancer.

The circular object appearing to exist as something in space is in fact a mere *appearance* existing only *within* the fire flow created by the dancer. If the rotation stops, the circle no more appears. Notice that the circle never existed to begin with; it was just a ball of fire only appearing to be a circle due to some constant flux.

Now if you hit such a circle with a stick, it will lose its shape and disappear altogether, because you disturbed the very flow *within* which it existed as appearance. This is very similar to what happens in the quantum world when measurement is performed. The act of measurement destroys the very appearance under investigation.

We see that some of the strange features of quantum physics are not so strange if we correct our classical view of the world by deeply reflecting on experiences within which we *have* the world in the first place. If objective knowledge has to begin with experience, end with experience, and never lose its rootedness in experience, then our slightest ignorance about the vital role of experience in the constitution of reality leads to a knowledge detached from reality.

The facts of the microscopic world articulated in quantum physics are experimental references to the underlying reality as a Heraclitean Flux, which is essentially a temporal-subjective reality rather than a spatial-objective theoretical fiction.

Nondual Reality

The nondual reality of Advaita Vedanta metaphysics and Transcendental Phenomenology can be fully expressed using the language of vector algebra as in quantum physics. It is precisely due to their consensus about the character of ultimate reality that they can all be expressed using one mathematical language.

I use this language to express the central idea in Advaita Vedanta metaphysics. It is not necessary to do the same for Transcendental Phenomenology because they are almost identical formulations of the life of pure consciousness. I personally view Transcendental Phenomenology as a modern embodiment of Advaita Vedanta metaphysics.

The one Brahman, pure consciousness, is the only thing that exists; it is nondual and admits no distinctions. All distinctions are appearances (reflections) of the same Brahman in relative planes of existence created by the mind. Brahman, being itself unaffected by its infinite manifestations, contains at once all its possibilities, just as a vector ontologically contains all its possible representations and components while being always above and beyond them.

The components *potentially* contained in the vector cannot be said to exist or not. Components have relative, dependent existence, while the vector itself has absolute, independent existence. Although components represent the vector, they are not identical with the vector; yet the vector and its components are one. In the same manner, Brahman and its manifestations are one.

In Advaita Vedanta metaphysics, the principle, pure consciousness, and its manifestations are related in exactly the same way that a vector is related to its representations.

Unity

In the language of quantum physics, the world we experience is really an *eigenstate* and not the world in itself. "World in itself" is a pure potentiality; it is an idea: it is something not yet differentiated into this and that, you and I, us and them, west and east, friend and enemy, man and woman, good and evil, etc.

Objects as distinct and separate entities exist only in names and forms. This separation has no substantial reality; it is itself an indivisible appearance within a primordial flux called *time*.

The underlying reality, which is a superposition state, a pure potentiality, does not lend itself to distinction; it is essentially nondual. Separation and isolation are effects that we produce ourselves in this unified reality by our instruments, which stand for our agendas, which are already intended and designed to perceive separation and division by creating it.

If the instruments of our Western sciences have come to a halt in their insatiable appetite for dividing and conquering, there may be a lesson for all of us in quantum physics, especially because its revelation coincided with an age of tremendous violence and destruction:

Quantum physics is reflective of nature's inherent resistance to separation, to being thrown into isolation and turned into a mere object of investigation. Nature likes communion but dislikes dominion.

The lesson we can learn from the Western science of physics is about the collapse of the policy of *divide and conquer* systematically adopted by this same West in all its projects, whether in science, philosophy, or politics.

If the wild West after centuries of intense inquiries has come to a conclusion already known in the East from thousands of years ago, we are by no means encouraged to become obsessed with the East and damn the West. It is time for this hypnotized world to transcend the East and the West and face up to the perennial unity awaiting its awakening.

Unity is always presupposed in diversity; it is only in the background of unity that diversity makes any sense. This background unity, that which stays the same throughout all apparent changes, is the one experiencing consciousness.

Without such transcendental unity in consciousness, we could not experience this phenomenal world as one world that is there for all of us.

If we look out into the world and cannot find the lost paradise, it is because *we are the lost paradise*. This witnessing, experiencing consciousness objectifies the world and itself into a being *inside* the world.

This world is nothing but the reflection of the ideas we entertain. We will perceive only diversity and destruction if we lose touch with human ideals, with a sense of the sacred, and with roots inside a consciousness that connects us from within.

One important conclusion shared by quantum physics, Advaita Vedanta, and Phenomenology is that *what we find in the world, we ourselves put there.*

Advaita Vedanta is an expression of the naked truth at the center of a circle on whose perimeter stand all mystics and esoteric traditions pointing inward. We are all pointing inward toward the same point at the center, for we are all outward projections of that same point. It belongs to our nature that we all want to go home.

Transcendental Phenomenology is the climax of human intellectual rigor in its search for truth. Phenomenology is not just a philosophy: it is a *way*, an esoteric system of mathematical precision that takes the seeker to the ground of the Absolute and the Infinite Truth by a bold tearing apart of all veils. Transcendental Phenomenology is the quintessence of all esoteric traditions suitable for the intellectually oriented seeker of truth. Phenomenology begins with the self and arrives at the self without ever leaving the sphere of lived experience. Phenomenology's contribution to the body of gnostic truths is that *the seeker is both the veil and the man behind the veil.*

Quantum physics tells us more about the macroscopic world than the microscopic world. It is a hard blow to our traditional-classical, rigid modes of thinking, which are phenomenologically blind to their own vital role in the constitution of the world and its destiny. Quantum physics is an expression of the breaking point of the *old world and its concepts and methods*. One can and one should without hasty and superficial judgment gain from this discipline both intellectually and spiritually as much as we have gained from it technologically.

If the world as depicted by our most valid science of today shares a deep truth with the world as pictured by thousands of mystics, then perhaps the idea of *truth* wasn't just a survival tool but a calling from our true essence for us to fully realize it.

In our search for the underlying truth of existence, we put behind us the age of religion and spirituality and adopted reason and science as our guides. Now we are at the verge of leaving behind the age in which our existential identities were fed by the theoretical models of sciences.

While spirituality and science remain, and will always remain, operative with respect to their originally intended purpose of providing comfort and happiness, we are entering the postspiritual era in which we become closer than ever to being awakened to ourselves. It is a postspiritual phase of consciousness because the vision of truth and liberation is no more tied to the notions of faith and unquestioned devotion but, on the contrary, to the freedom from all faith and conviction, and therefore seeing the truth for ourselves.

It is no more up to the elitist traditions and gurus and scientists to preserve the truth for a select few who should undergo initiation and years of formal training in order to see *the face of truth*. Truth is not something to be earned by unquestioned submission to the *other*; truth is something to be remembered by turning toward the *self*.

The truth, the absolute and the infinite truth that we are destined to face, is not outside us; it is not inside us; yet it is the closest thing to us, because it is both us and not us. Truth is the face of the Self concealed by the self.

Let us conclude the book by a mystical hint of Isha Upanishad in the direction of that perennial truth whose sight is its first and final proof:

The face of truth is concealed by a golden vessel.

Printed in Great Britain
by Amazon

61913782R00054